Beekeeping A Practicle Guide that Supplies Business Book Knowledge for Beginners
How to Start, Finance & Market a Beekeeping Business

by

Brian Mahoney

Get Our Video Training Program at:

(Zero Cost Internet Marketing complete 142 video series)

http://goo.gl/gQnSo4

Massive Money for Real Estate Investing

http://www.BrianSMahoney.com

ABOUT THE AUTHOR

Brian Mahoney is the author of over 100 business start-up guides, real estate investing programs and Christian literature. He started his company MahoneyProducts in 1992.

He served in the US Army and worked over a decade for the US Postal Service. An active real estate investor, he has also served as a minister for the Churches of Christ in Virginia and Michigan.

He has degree's in Business Administration and Applied Science & Computer Programming.

His books and video training programs have helped thousands of people all over the world start there own successful business.

http://www.briansmahoney.com/

Copyright © 2016 Brian Mahoney
All rights reserved.

DEDICATION

**This book is dedicated to my son's
Christian and Matthew.
A blessing from God and the joy of my life.**

Table of Contents

Chapter 1 Business Overview

Chapter 2 A Step by Step Plan

Chapter 3 Business Plan

Chapter 4 Free Money to Get Started

Chapter 5 SBA Loans

Chapter 6 Micro Loans

Chapter 7 Legal Structure

Chapter 8 Business Name

Chapter 9 Business Tax Advantages

Chapter 10 How to Hire Employees

Chapter 11 Million Dollar Wholesale Rolodex

Chapter 12 Zero Cost Marketing

Chapter 13 Bonus Material: Motivation

Chapter 14 Bonus Material: Credit Repair

ACKNOWLEDGMENTS

I WOULD LIKE TO ACKNOWLEDGE ALL THE HARD WORK OF THE MEN AND WOMEN OF THE UNITED STATES MILITARY, WHO RISK THEIR LIVES ON A DAILY BASIS, TO MAKE THE WORLD A SAFER PLACE.

Disclaimer

This book was written as a guide to starting a business. As with any other high yielding action, starting a business has a certain degree of risk. This book is not meant to take the place of accounting, legal, financial or other professional advice. If advice is needed in any of these fields, you are advised to seek the services of a professional.

While the author has attempted to make the information in this book as accurate as possible, no guarantee is given as to the accuracy or currency of any individual item. Laws and procedures related to business are constantly changing.

Therefore, in no event shall Brian Mahoney, the author of this book be liable for any special, indirect, or consequential damages or any damages whatsoever in connection with the use of the information herein provided.

All Rights Reserved

No part of this book may be used or reproduced in any manner whatsoever without the written permission of the author.

Beekeeping

Apiculture (from Latin: apis "bee") is the maintenance of honey bee colonies, commonly in hives, by humans. A beekeeper (or apiarist) keeps bees in order to collect their honey and other products that the hive produces (including beeswax, propolis, pollen, and royal jelly), to pollinate crops, or to produce bees for sale to other beekeepers. A location where bees are kept is called an apiary or "bee yard".

Depictions of humans collecting honey from wild bees date to 15,000 years ago. Beekeeping in pottery vessels began about 9,000 years ago in North Africa. Domestication is shown in Egyptian art from around 4,500 years ago. Simple hives and smoke were used and honey was stored in jars, some of which were found in the tombs of pharaohs such as Tutankhamun. It wasn't until the 18th century that European understanding of the colonies and biology of bees allowed the construction of the moveable comb hive so that honey could be harvested without destroying the entire colony.

History of beekeeping

At some point humans began to attempt to domesticate wild bees in artificial hives made from hollow logs, wooden boxes, pottery vessels, and woven straw baskets or "skeps". Traces of beeswax are found in pot sherds throughout the Middle East beginning about 7000 BCE.

Honeybees were kept in Egypt from antiquity. On the walls of the sun temple of Nyuserre Ini from the Fifth Dynasty, before 2422 BCE, workers are depicted blowing smoke into hives as they are removing honeycombs. Inscriptions detailing the production of honey are found on the tomb of Pabasa from the Twenty-sixth Dynasty (c. 650 BCE), depicting pouring honey in jars and cylindrical hives. Sealed pots of honey were found in the grave goods of pharaohs such as Tutankhamun.

There was a documented attempt to introduce bees to dry areas of Mesopotamia in the 8th century BCE by Shamash-resh-uṣur, the governor of Mari and Suhu. His plans were detailed in a stele of 760 BCE:

I am Shamash-resh-uṣur , the governor of Suhu and the land of Mari. Bees that collect honey, which none of my ancestors had ever seen or brought into the land of Suhu, I brought down from the mountain of the men of Habha, and made them settle in the orchards of the town 'Gabbari-built-it'. They collect honey and wax, and I know how to melt the honey and wax – and the gardeners know too. Whoever comes in the future, may he ask the old men of the town, (who will say) thus: "They are the buildings of Shamash-resh-uṣur, the governor of Suhu, who introduced honey bees into the land of Suhu."

—translated text from stele, (Dalley, 2002)

In prehistoric Greece (Crete and Mycenae), there existed a system of high-status apiculture, as can be concluded from the finds of hives, smoking pots, honey extractors and other beekeeping paraphernalia in Knossos. Beekeeping was considered a highly valued industry controlled by beekeeping overseers—owners of gold rings depicting apiculture scenes rather than religious ones as they have been reinterpreted recently, contra Sir Arthur Evans.

Archaeological finds relating to beekeeping have been discovered at Rehov, a Bronze and Iron Age archaeological site in the Jordan Valley, Israel. Thirty intact hives, made of straw and unbaked clay, were discovered by archaeologist Amihai Mazar in the ruins of the city, dating from about 900 BCE. The hives were found in orderly rows, three high, in a manner that could have accommodated around 100 hives, held more than 1 million bees and had a potential annual yield of 500 kilograms of honey and 70 kilograms of beeswax, according to Mazar, and are evidence that an advanced honey industry existed in ancient Israel 3,000 years ago.

In ancient Greece, aspects of the lives of bees and beekeeping are discussed at length by Aristotle. Beekeeping was also documented by the Roman writers Virgil, Gaius Julius Hyginus, Varro, and Columella.

Beekeeping has also been practiced in ancient China since antiquity. In the book "Golden Rules of Business Success" written by Fan Li (or Tao Zhu Gong) during the Spring and Autumn Period there are sections describing the art of beekeeping, stressing the importance of the quality of the wooden box used and how this can affect the quality of the honey.

The ancient Maya domesticated a separate species of stingless bee. The use of stingless bees is referred to as meliponiculture, named after bees of the tribe Meliponini —such as Melipona quadrifasciata in Brazil. This variation of bee keeping still occurs around the world today. For instance, in Australia, the stingless bee Tetragonula carbonaria is kept for production of their honey.

Origins

There are more than 20,000 species of wild bees. Many species are solitary (e.g., mason bees, leafcutter bees (Megachilidae), carpenter bees and other ground-nesting bees). Many others rear their young in burrows and small colonies (e.g., bumblebees and stingless bees). Some honey bees are wild e.g. the little honeybee (Apis florea), giant honeybee (Apis dorsata) and rock bee (Apis laboriosa). Beekeeping, or apiculture, is concerned with the practical management of the social species of honey bees, which live in large colonies of up to 100,000 individuals.

In Europe and America the species universally managed by beekeepers is the Western honey bee (Apis mellifera). This species has several sub-species or regional varieties, such as the Italian bee (Apis mellifera ligustica), European dark bee (Apis mellifera mellifera), and the Carniolan honey bee (Apis mellifera carnica). In the tropics, other species of social bees are managed for honey production, including the Asiatic honey bee (Apis cerana).

All of the Apis mellifera sub-species are capable of inter-breeding and hybridizing. Many bee breeding companies strive to selectively breed and hybridize varieties to produce desirable qualities: disease and parasite resistance, good honey production, swarming behaviour reduction, prolific breeding, and mild disposition. Some of these hybrids are marketed under specific brand names, such as the Buckfast Bee or Midnite Bee. The advantages of the initial F1 hybrids produced by these crosses include: hybrid vigor, increased honey productivity, and greater disease resistance. The disadvantage is that in subsequent generations these advantages may fade away and hybrids tend to be very defensive and aggressive.

Wild honey harvesting

Collecting honey from wild bee colonies is one of the most ancient human activities and is still practiced by aboriginal societies in parts of Africa, Asia, Australia, and South America. In Africa, honeyguide birds have evolved a mutualist relationship with humans, leading them to hives and participating in the feast. This suggests honey harvesting by humans may be of great antiquity. Some of the earliest evidence of gathering honey from wild colonies is from rock paintings, dating to around Upper Paleolithic (13,000 BCE). Gathering honey from wild bee colonies is usually done by subduing the bees with smoke and breaking open the tree or rocks where the colony is located, often resulting in the physical destruction of the nest.

Study of honey bees

It was not until the 18th century that European natural philosophers undertook the scientific study of bee colonies and began to understand the complex and hidden world of bee biology. Preeminent among these scientific pioneers were Swammerdam, René Antoine Ferchault de Réaumur, Charles Bonnet, and Francois Huber. Swammerdam and Réaumur were among the first to use a microscope and dissection to understand the internal biology of honey bees.

Réaumur was among the first to construct a glass walled observation hive to better observe activities within hives. He observed queens laying eggs in open cells, but still had no idea of how a queen was fertilized; nobody had ever witnessed the mating of a queen and drone and many theories held that queens were "self-fertile," while others believed that a vapor or "miasma" emanating from the drones fertilized queens without direct physical contact. Huber was the first to prove by observation and experiment that queens are physically inseminated by drones outside the confines of hives, usually a great distance away.

Following Réaumur's design, Huber built improved glass-walled observation hives and sectional hives that could be opened like the leaves of a book. This allowed inspecting individual wax combs and greatly improved direct observation of hive activity. Although he went blind before he was twenty, Huber employed a secretary, Francois Burnens, to make daily observations, conduct careful experiments, and keep accurate notes over more than twenty years. Huber confirmed that a hive consists of one queen who is the mother of all the female workers and male drones in the colony. He was also the first to confirm that mating with drones takes place outside of hives and that queens are inseminated by a number of successive matings with male drones, high in the air at a great distance from their hive.

Together, he and Burnens dissected bees under the microscope and were among the first to describe the ovaries and spermatheca, or sperm store, of queens as well as the penis of male drones. Huber is universally regarded as "the father of modern bee-science" and his "Nouvelles Observations sur Les Abeilles (or "New Observations on Bees") revealed all the basic scientific truths for the biology and ecology of honeybees.

Rural beekeeping in the 16th century

Early forms of honey collecting entailed the destruction of the entire colony when the honey was harvested. The wild hive was crudely broken into, using smoke to suppress the bees, the honeycombs were torn out and smashed up — along with the eggs, larvae and honey they contained. The liquid honey from the destroyed brood nest was strained through a sieve or basket. This was destructive and unhygienic, but for hunter-gatherer societies this did not matter, since the honey was generally consumed immediately and there were always more wild colonies to exploit. But in settled societies the destruction of the bee colony meant the loss of a valuable resource; this drawback made beekeeping both inefficient and something of a "stop and start" activity. There could be no continuity of production and no possibility of selective breeding, since each bee colony was destroyed at harvest time, along with its precious queen.

During the medieval period abbeys and monasteries were centers of beekeeping, since beeswax was highly prized for candles and fermented honey was used to make alcoholic mead in areas of Europe where vines would not grow. The 18th and 19th centuries saw successive stages of a revolution in beekeeping, which allowed the bees themselves to be preserved when taking the harvest.

Intermediate stages in the transition from the old beekeeping to the new were recorded for example by Thomas Wildman in 1768/1770, who described advances over the destructive old skep-based beekeeping so that the bees no longer had to be killed to harvest the honey. Wildman for example fixed a parallel array of wooden bars across the top of a straw hive or skep (with a separate straw top to be fixed on later) "so that there are in all seven bars of deal" [in a 10-inch-diameter (250 mm) hive] "to which the bees fix their combs." He also described using such hives in a multi-storey configuration, foreshadowing the modern use of supers: he described adding (at a proper time) successive straw hives below, and eventually removing the ones above when free of brood and filled with honey, so that the bees could be separately preserved at the harvest for a following season. Wildman also described a further development, using hives with "sliding frames" for the bees to build their comb, foreshadowing more modern uses of movable-comb hives.

Wildman's book acknowledged the advances in knowledge of bees previously made by Swammerdam, Maraldi, and de Réaumur—he included a lengthy translation of Réaumur's account of the natural history of bees—and he also described the initiatives of others in designing hives for the preservation of bee-life when taking the harvest, citing in particular reports from Brittany dating from the 1750s, due to Comte de la Bourdonnaye. However, the forerunners of the modern hives with movable frames that are mainly used today are considered the traditional basket top bar (movable comb) hives of Greece, known as "Greek beehives". The oldest testimony on their use dates back to 1669 although it is probable that their use is more than 3000 years old.

The 19th century saw this revolution in beekeeping practice completed through the perfection of the movable comb hive by the American Lorenzo Lorraine Langstroth. Langstroth was the first person to make practical use of Huber's earlier discovery that there was a specific spatial measurement between the wax combs, later called the bee space, which bees do not block with wax, but keep as a free passage. Having determined this bee space (between 5 and 8 mm, or 1/4 to 3/8"), Langstroth then designed a series of wooden frames within a rectangular hive box, carefully maintaining the correct space between successive frames, and found that the bees would build parallel honeycombs in the box without bonding them to each other or to the hive walls. This enables the beekeeper to slide any frame out of the hive for inspection, without harming the bees or the comb, protecting the eggs, larvae and pupae contained within the cells. It also meant that combs containing honey could be gently removed and the honey extracted without destroying the comb. The emptied honey combs could then be returned to the bees intact for refilling. Langstroth's book, The Hive and Honey-bee, published in 1853, described his rediscovery of the bee space and the development of his patent movable comb hive.

The invention and development of the movable-comb-hive fostered the growth of commercial honey production on a large scale in both Europe and the USA (see also Beekeeping in the United States).

Evolution of hive designs

Langstroth's design for movable comb hives was seized upon by apiarists and inventors on both sides of the Atlantic and a wide range of moveable comb hives were designed and perfected in England, France, Germany and the United States. Classic designs evolved in each country: Dadant hives and Langstroth hives are still dominant in the USA; in France the De-Layens trough-hive became popular and in the UK a British National hive became standard as late as the 1930s although in Scotland the smaller Smith hive is still popular. In some Scandinavian countries and in Russia the traditional trough hive persisted until late in the 20th century and is still kept in some areas. However, the Langstroth and Dadant designs remain ubiquitous in the USA and also in many parts of Europe, though Sweden, Denmark, Germany, France and Italy all have their own national hive designs. Regional variations of hive evolved to reflect the climate, floral productivity and the reproductive characteristics of the various subspecies of native honey bee in each bio-region.

The differences in hive dimensions are insignificant in comparison to the common factors in all these hives: they are all square or rectangular; they all use movable wooden frames; they all consist of a floor, brood-box, honey super, crown-board and roof. Hives have traditionally been constructed of cedar, pine, or cypress wood, but in recent years hives made from injection molded dense polystyrene have become increasingly important.

Hives also use queen excluders between the brood-box and honey supers to keep the queen from laying eggs in cells next to those containing honey intended for consumption. Also, with the advent in the 20th century of mite pests, hive floors are often replaced for part of (or the whole) year with a wire mesh and removable tray.

Pioneers of practical and commercial beekeeping

The 19th century produced an explosion of innovators and inventors who perfected the design and production of beehives, systems of management and husbandry, stock improvement by selective breeding, honey extraction and marketing. Preeminent among these innovators were:

Petro Prokopovych, used frames with channels in the side of the woodwork, these were packed side by side in boxes that were stacked one on top of the other. The bees travelling from frame to frame and box to box via the channels. The channels were similar to the cut outs in the sides of modern wooden sections (1814).

Jan Dzierżon, was the father of modern apiology and apiculture. All modern beehives are descendants of his design.

L. L. Langstroth, revered as the "father of American apiculture", no other individual has influenced modern beekeeping practice more than Lorenzo Lorraine Langstroth. His classic book The Hive and Honey-bee was published in 1853.

Moses Quinby, often termed 'the father of commercial beekeeping in the United States', author of Mysteries of Bee-Keeping Explained.

Amos Root, author of the A B C of Bee Culture, which has been continuously revised and remains in print. Root pioneered the manufacture of hives and the distribution of bee-packages in the United States.

A. J. Cook, author of The Bee-Keepers' Guide; or Manual of the Apiary, 1876.

Dr. C.C. Miller was one of the first entrepreneurs to actually make a living from apiculture. By 1878 he made beekeeping his sole business activity. His book, Fifty Years Among the Bees, remains a classic and his influence on bee management persists to this day.

Honey spinner

Major Francesco De Hruschka was an Italian military officer who made one crucial invention that catalyzed the commercial honey industry. In 1865 he invented a simple machine for extracting honey from the comb by means of centrifugal force. His original idea was simply to support combs in a metal framework and then spin them around within a container to collect honey as it was thrown out by centrifugal force. This meant that honeycombs could be returned to a hive undamaged but empty, saving the bees a vast amount of work, time, and materials. This single invention greatly improved the efficiency of honey harvesting and catalysed the modern honey industry.

Walter T. Kelley was an American pioneer of modern beekeeping in the early and mid-20th century. He greatly improved upon beekeeping equipment and clothing and went on to manufacture these items as well as other equipment. His company sold via catalog worldwide and his book, How to Keep Bees & Sell Honey, an introductory book of apiculture and marketing, allowed for a boom in beekeeping following World War II.

In the U.K. practical beekeeping was led in the early 20th century by a few men, pre-eminently Brother Adam and his Buckfast bee and R.O.B. Manley, author of many titles, including Honey Production in the British Isles and inventor of the Manley frame, still universally popular in the U.K. Other notable British pioneers include William Herrod-Hempsall and Gale.

Dr. Ahmed Zaky Abushady (1892–1955), was an Egyptian poet, medical doctor, bacteriologist and bee scientist who was active in England and in Egypt in the early part of the twentieth century. In 1919, Abushady patented a removable, standardized aluminum honeycomb. In 1919 he also founded The Apis Club in Benson, Oxfordshire, and its periodical Bee World, which was to be edited by Annie D. Betts and later by Dr. Eva Crane.

The Apis Club was transitioned to the International Bee Research Association (IBRA). Its archives are held in the National Library of Wales. In Egypt in the 1930s, Abushady established The Bee Kingdom League and its organ, The Bee Kingdom.

In India, R. N. Mattoo was the pioneer worker in starting beekeeping with Indian honeybee, (Apis cerana indica) in early 1930s. Beekeeping with European honeybee, (Apis mellifera) was started by Dr. A. S. Atwal and his team members, O. P. Sharma and N. P. Goyal in Punjab in early 1960s. It remained confined to Punjab and Himachal Pradesh up to late 1970s. Later on in 1982, Dr. R. C. Sihag, working at Haryana Agricultural University, Hisar (Haryana), introduced and established this honeybee in Haryana and standardized its management practices for semi-arid-subtropical climates. On the basis of these practices, beekeeping with this honeybee could be extended to the rest of the country. Now beekeeping with Apis mellifera predominates in India.

Traditional beekeeping

A fixed comb hive is a hive in which the combs cannot be removed or manipulated for management or harvesting without permanently damaging the comb. Almost any hollow structure can be used for this purpose, such as a log gum, skep, wooden box, or a clay pot or tube.

Fixed comb hives are no longer in common use in industrialized countries, and are illegal in places that require movable combs to inspect for problems such as varroa and American foulbrood. In many developing countries fixed comb hives are widely used and, because they can be made from any locally available material, are very inexpensive. Beekeeping using fixed comb hives is an essential part of the livelihoods of many communities in poor countries. The charity Bees for Development recognizes that local skills to manage bees in fixed comb hives are widespread in Africa, Asia, and South America. Internal size of fixed comb hives range from 32.7 liters (2000 cubic inches) typical of the clay tube hives used in Egypt to 282 liters (17209 cubic inches) for the Perone hive. Straw skeps, bee gums, and unframed box hives are unlawful in most US states, as the comb and brood cannot be inspected for diseases. However, skeps are still used for collecting swarms by hobbyists in the UK, before moving them into standard hives. Quinby used box hives to produce so much honey that he saturated the New York market in the 1860s. His writings contain excellent advice for management of bees in fixed comb hives.

Modern beekeeping

Top bar hives have been widely adopted in Africa where they are used to keep tropical honeybee ecotypes. Their advantages include being light weight, adaptable, easy to harvest honey, and less stressful for the bees. Disadvantages include combs that are fragile and cannot usually be extracted and returned to the bees to be refilled and that they cannot easily be expanded for additional honey storage.

A growing number of amateur beekeepers are adopting various top-bar hives similar to the type commonly found in Africa. Top bar hives were originally used as a traditional beekeeping method in Greece and Vietnam with a history dating back over 2000 years. These hives have no frames and the honey-filled comb is not returned after extraction. Because of this, the production of honey is likely to be somewhat less than that of a frame and super based hive such as Langstroth or Dadant. Top bar hives are mostly kept by people who are more interested in having bees in their garden than in honey production per se. Some of the most well known top-bar hive designs are the Kenyan Top Bar Hive with sloping sides, the Tanzanian Top Bar Hive with straight sides, and Vertical Top Bar Hives, such as the Warre or "People's Hive" designed by Abbe Warre in the mid-1900s.

The initial costs and equipment requirements are typically much less than other hive designs. Scrap wood or #2 or #3 pine can often be used to build a nice hive. Top-bar hives also offer some advantages to interacting with the bees and the amount of weight that must be lifted is greatly reduced. Top-bar hives are being widely used in developing countries in Africa and Asia as a result of the Bees for Development program. Since 2011, a growing number of beekeepers in the U.S. are using various top-bar hives.

Horizontal frame hives

The De-Layens hive, Jackson Horizontal Hive, and various chest type hives are widely used in Spain, France, Ukraine, Belarus, Africa, and parts of Russia. They are a step up from fixed comb and top bar hives because they have movable frames that can be extracted. Their limitation is primarily that volume is fixed and not easily expanded. Honey has to be removed one frame at a time, extracted or crushed, and the empty frames returned to be refilled. Various horizontal hives have been adapted and widely used for commercial migratory beekeeping. The Jackson Horizontal Hive is particularly well adapted for tropical agriculture. The De-Layens hive is popular in parts of Spain.

Vertical stackable frame hives

In the United States, the Langstroth hive is commonly used. The Langstroth was the first successful top-opened hive with movable frames. Many other hive designs are based on the principle of bee space first described by Langstroth. The Langstroth hive is a descendant of Jan Dzierzon's Polish hive designs. In the United Kingdom, the most common type of hive is the British National, which can hold Hoffman, British Standard or Manley frames. It is not unusual to see some other sorts of hive (Smith, Commercial, WBC, Langstroth, and Rose). Dadant and Modified Dadant hives are widely used in France and Italy where their large size is an advantage. Square Dadant hives - often called 12 frame Dadant or Brother Adam hives - are used in large parts of Germany and other parts of Europe by commercial beekeepers. The Rose hive is a modern design that attempts to address many of the flaws and limitations of other movable frame hives. The only significant weakness of the Rose design is that it requires 2 or 3 boxes as a brood nest which infers a large number of frames to be worked when managing the bees. The major advantage shared by these designs is that additional brood and honey storage space can be added via boxes of frames added to the hive. This also simplifies honey collection since an entire box of honey can be removed instead of removing one frame at a time.

Protective clothing

Most beekeepers also wear some protective clothing. Novice beekeepers usually wear gloves and a hooded suit or hat and veil. Experienced beekeepers sometimes elect not to use gloves because they inhibit delicate manipulations. The face and neck are the most important areas to protect, so most beekeepers wear at least a veil. Defensive bees are attracted to the breath, and a sting on the face can lead to much more pain and swelling than a sting elsewhere, while a sting on a bare hand can usually be quickly removed by fingernail scrape to reduce the amount of venom injected.

The protective clothing is generally light colored (but not colorful) and of a smooth material. This provides the maximum differentiation from the colony's natural predators (such as bears and skunks) which tend to be dark-colored and furry.

'Stings' retained in clothing fabric continue to pump out an alarm pheromone that attracts aggressive action and further stinging attacks. Washing suits regularly, and rinsing gloved hands in vinegar minimizes attraction.

Smoker

Smoke is the beekeeper's third line of defense. Most beekeepers use a "smoker"—a device designed to generate smoke from the incomplete combustion of various fuels. Smoke calms bees; it initiates a feeding response in anticipation of possible hive abandonment due to fire. Smoke also masks alarm pheromones released by guard bees or when bees are squashed in an inspection. The ensuing confusion creates an opportunity for the beekeeper to open the hive and work without triggering a defensive reaction. In addition, when a bee consumes honey the bee's abdomen distends, supposedly making it difficult to make the necessary flexes to sting, though this has not been tested scientifically.

Smoke is of questionable use with a swarm, because swarms do not have honey stores to feed on in response. Usually smoke is not needed, since swarms tend to be less defensive, as they have no stores or brood to defend, and a fresh swarm has fed well from the hive.

Many types of fuel can be used in a smoker as long as it is natural and not contaminated with harmful substances. These fuels include hessian, twine, burlap, pine needles, corrugated cardboard, and mostly rotten or punky wood. Indian beekeepers, especially in Kerala, often use coconut fibers as they are readily available, safe, and of negligible expense. Some beekeeping supply sources also sell commercial fuels like pulped paper and compressed cotton, or even aerosol cans of smoke. Other beekeepers use sumac as fuel because it ejects lots of smoke and doesn't have an odor.

Some beekeepers are using "liquid smoke" as a safer, more convenient alternative. It is a water-based solution that is sprayed onto the bees from a plastic spray bottle.

Torpor may also be induced by the introduction of chilled air into the hive – while chilled carbon dioxide may have harmful long-term effects.

Effects of stings and of protective measures

Some beekeepers believe that the more stings a beekeeper receives, the less irritation each causes, and they consider it important for safety of the beekeeper to be stung a few times a season. Beekeepers have high levels of antibodies (mainly IgG) reacting to the major antigen of bee venom, phospholipase A2 (PLA). Antibodies correlate with the frequency of bee stings.

The entry of venom into the body from bee-stings may also be hindered and reduced by protective clothing that allows the wearer to remove stings and venom sacs with a simple tug on the clothing. Although the stinger is barbed, a worker bee is less likely to become lodged into clothing than human skin.

If a beekeeper is stung by a bee, there are many protective measures that should be taken in order to make sure the affected area does not become too irritated. The first cautionary step that should be taken following a bee sting is removing the stinger without squeezing the attached venom glands. A quick scrape with a fingernail is effective and intuitive. This step is effective in making sure that the venom injected does not spread, so the side effects of the sting will go away sooner. Washing the affected area with soap and water is also a good way to stop the spread of venom. The last step that needs to be taken is to apply ice or a cold compress to the stung area.

Natural beekeeping

The natural beekeeping movement believes that modern beekeeping and agricultural practices, such as crop spraying, hive movement, frequent hive inspections, artificial insemination of queens, routine medication, and sugar water feeding, weaken bee hives.

Practitioners of 'natural beekeeping' tend to use variations of the top-bar hive, which is a simple design that retains the concept of movable comb without the use of frames or foundation. The horizontal top-bar hive, as championed by Marty Hardison, Michael Bush, Philip Chandler, Dennis Murrell and others, can be seen as a modernization of hollow log hives, with the addition of wooden bars of specific width from which bees hang their combs. Its widespread adoption in recent years can be attributed to the publication in 2007 of The Barefoot Beekeeper by Philip Chandler, which challenged many aspects of modern beekeeping and offered the horizontal top-bar hive as a viable alternative to the ubiquitous Langstroth-style movable-frame hive.

The most popular vertical top-bar hive is probably the Warré hive, based on a design by the French priest Abbé Émile Warré (1867–1951) and popularized by Dr. David Heaf in his English translation of Warré's book L'Apiculture pour Tous as Beekeeping For All.

Urban or backyard beekeeping

Related to natural beekeeping, urban beekeeping is an attempt to revert to a less industrialized way of obtaining honey by utilizing small-scale colonies that pollinate urban gardens. Urban apiculture has undergone a renaissance in the first decade of the 21st century, and urban beekeeping is seen by many as a growing trend.

Some have found that "city bees" are actually healthier than "rural bees" because there are fewer pesticides and greater biodiversity. Urban bees may fail to find forage, however, and homeowners can use their landscapes to help feed local bee populations by planting flowers that provide nectar and pollen. An environment of year-round, uninterrupted bloom creates an ideal environment for colony reproduction.

Bee colonies

A colony of bees consists of three castes of bee:

    a queen bee, which is normally the only breeding female in the colony;

a large number of female worker bees, typically 30,000–50,000 in number;

a number of male drones, ranging from thousands in a strong hive in spring to very few during dearth or cold season.

Queen bee

The queen is the only sexually mature female in the hive and all of the female worker bees and male drones are her offspring. The queen may live for up to three years or more and may be capable of laying half a million eggs or more in her lifetime. At the peak of the breeding season, late spring to summer, a good queen may be capable of laying 3,000 eggs in one day, more than her own body weight. This would be exceptional however; a prolific queen might peak at 2,000 eggs a day, but a more average queen might lay just 1,500 eggs per day. The queen is raised from a normal worker egg, but is fed a larger amount of royal jelly than a normal worker bee, resulting in a radically different growth and metamorphosis. The queen influences the colony by the production and dissemination of a variety of pheromones or "queen substances". One of these chemicals suppresses the development of ovaries in all the female worker bees in the hive and prevents them from laying eggs.

Mating of queens

The queen emerges from her cell after 15 days of development and she remains in the hive for 3–7 days before venturing out on a mating flight. Mating flight is otherwise known as 'nuptial flight'. Her first orientation flight may only last a few seconds, just enough to mark the position of the hive. Subsequent mating flights may last from 5 minutes to 30 minutes, and she may mate with a number of male drones on each flight. Over several matings, possibly a dozen or more, the queen receives and stores enough sperm from a succession of drones to fertilize hundreds of thousands of eggs. If she does not manage to leave the hive to mate—possibly due to bad weather or being trapped in part of the hive—she remains infertile and become a drone layer, incapable of producing female worker bees. Worker bees sometimes kill a non-performing queen and produce another. Without a properly performing queen, the hive is doomed.

Mating takes place at some distance from the hive and often several hundred feet in the air; it is thought that this separates the strongest drones from the weaker ones, ensuring that only the fastest and strongest drones get to pass on their genes.

Worker bees

Female worker bee

Almost all the bees in a hive are female worker bees. At the height of summer when activity in the hive is frantic and work goes on non-stop, the life of a worker bee may be as short as 6 weeks; in late autumn, when no brood is being raised and no nectar is being harvested, a young bee may live for 16 weeks, right through the winter.

Over the course of their lives, worker bees' duties are dictated by age. For the first few weeks of their lifespan, they perform basic chores within the hive: cleaning empty brood cells, removing debris and other housekeeping tasks, making wax for building or repairing comb, and feeding larvae. Later, they may ventilate the hive or guard the entrance. Older workers leave the hive daily, weather permitting, to forage for nectar, pollen, water, and propolis.

Period	Work activity
Days 1-3	Cleaning cells and incubation
Day 3-6	Feeding older larvae
Day 6-10	Feeding younger larvae
Day 8-16	Receiving nectar and pollen from field bees
Day 12-18	Beeswax making and cell building
Day 14 onwards	Entrance guards; nectar, pollen, water and

Drones are the largest bees in the hive (except for the queen), at almost twice the size of a worker bee. They do not work, do not forage for pollen or nectar, are unable to sting, and have no other known function than to mate with new queens and fertilize them on their mating flights. A bee colony generally starts to raise drones a few weeks before building queen cells so they can supersede a failing queen or prepare for swarming. When queen-raising for the season is over, bees in colder climates drive drones out of the hive to die, biting and tearing their legs and wings.

Differing stages of development

Stage of development	Queen	Worker	Drone
Egg	3 days	3 days	3 days
Larva	8 days	10 days	13 days

:Successive moults occur within this period 8 to 13 day period

Cell Capped	day 8	day 8	day 10
Pupa	4 days	8 days	8 days
Total	15 days	21 days	24 days

Structure of a bee colony

A domesticated bee colony is normally housed in a rectangular hive body, within which eight to ten parallel frames house the vertical plates of honeycomb that contain the eggs, larvae, pupae and food for the colony. If one were to cut a vertical cross-section through the hive from side to side, the brood nest would appear as a roughly ovoid ball spanning 5-8 frames of comb. The two outside combs at each side of the hive tend to be exclusively used for long-term storage of honey and pollen.

Within the central brood nest, a single frame of comb typically has a central disk of eggs, larvae and sealed brood cells that may extend almost to the edges of the frame. Immediately above the brood patch an arch of pollen-filled cells extends from side to side, and above that again a broader arch of honey-filled cells extends to the frame tops. The pollen is protein-rich food for developing larvae, while honey is also food but largely energy rich rather than protein rich. The nurse bees that care for the developing brood secrete a special food called 'royal jelly' after feeding themselves on honey and pollen. The amount of royal jelly fed to a larva determines whether it develops into a worker bee or a queen.

Apart from the honey stored within the central brood frames, the bees store surplus honey in combs above the brood nest. In modern hives the beekeeper places separate boxes, called 'supers', above the brood box, in which a series of shallower combs is provided for storage of honey. This enables the beekeeper to remove some of the supers in the late summer, and to extract the surplus honey harvest, without damaging the colony of bees and its brood nest below. If all the honey is 'stolen', including the amount of honey needed to survive winter, the beekeeper must replace these stores by feeding the bees sugar or corn syrup in autumn.

Annual cycle of a bee colony

The development of a bee colony follows an annual cycle of growth that begins in spring with a rapid expansion of the brood nest, as soon as pollen is available for feeding larvae. Some production of brood may begin as early as January, even in a cold winter, but breeding accelerates towards a peak in May (in the northern hemisphere), producing an abundance of harvesting bees synchronized to the main nectar flow in that region. Each race of bees times this build-up slightly differently, depending on how the flora of its original region blooms. Some regions of Europe have two nectar flows: one in late spring and another in late August. Other regions have only a single nectar flow. The skill of the beekeeper lies in predicting when the nectar flow will occur in his area and in trying to ensure that his colonies achieve a maximum population of harvesters at exactly the right time.

The key factor in this is the prevention or skillful management of the swarming impulse. If a colony swarms unexpectedly and the beekeeper does not manage to capture the resulting swarm, he is likely to harvest significantly less honey from that hive, since he has lost half his worker bees at a single stroke.

If, however, he can use the swarming impulse to breed a new queen but keep all the bees in the colony together, he maximizes his chances of a good harvest. It takes many years of learning and experience to be able to manage all these aspects successfully, though owing to variable circumstances many beginners often achieve a good honey harvest.

Formation of new colonies

All colonies are totally dependent on their queen, who is the only egg-layer. However, even the best queens live only a few years and one or two years longevity is the norm. She can choose whether or not to fertilize an egg as she lays it; if she does so, it develops into a female worker bee; if she lays an unfertilized egg it becomes a male drone. She decides which type of egg to lay depending on the size of the open brood cell she encounters on the comb. In a small worker cell, she lays a fertilized egg; if she finds a larger drone cell, she lays an unfertilized drone egg.

All the time that the queen is fertile and laying eggs she produces a variety of pheromones, which control the behavior of the bees in the hive. These are commonly called queen substance, but there are various pheromones with different functions.

As the queen ages, she begins to run out of stored sperm, and her pheromones begin to fail. Inevitably, the queen begins to falter, and the bees decide to replace her by creating a new queen from one of her worker eggs. They may do this because she has been damaged (lost a leg or an antenna), because she has run out of sperm and cannot lay fertilized eggs (has become a 'drone laying queen'), or because her pheromones have dwindled to where they cannot control all the bees in the hive.

At this juncture, the bees produce one or more queen cells by modifying existing worker cells that contain a normal female egg. However, the bees pursue two distinct behaviors:

Different sub-species of Apis mellifera exhibit differing swarming characteristics that reflect their evolution in different ecotopes of the European continent. In general the more northerly black races are said to swarm less and supersede more, whereas the more southerly yellow and grey varieties are said to swarm more frequently. The truth is complicated because of the prevalence of cross-breeding and hybridization of the sub species and opinions differ.

Supersedure is highly valued as a behavioral trait by beekeepers because a hive that supersedes its old queen does not swarm and so no stock is lost; it merely creates a new queen and allows the old one to fade away, or alternatively she is killed when the new queen emerges. When superseding a queen, the bees produce just one or two queen cells, characteristically in the center of the face of a broodcomb.

In swarming, by contrast, a great many queen cells are created—typically a dozen or more—and these are located around the edges of a broodcomb, most often at the sides and the bottom.

New wax combs between basement joists

Once either process has begun, the old queen normally leaves the hive with the hatching of the first queen cells. She leaves accompanied by a large number of bees, predominantly young bees (wax-secretors), who form the basis of the new hive. Scouts are sent out from the swarm to find suitable hollow trees or rock crevices. As soon as one is found, the entire swarm moves in. Within a matter of hours, they build new wax brood combs, using honey stores that the young bees have filled themselves with before leaving the old hive.

Only young bees can secrete wax from special abdominal segments, and this is why swarms tend to contain more young bees. Often a number of virgin queens accompany the first swarm (the 'prime swarm'), and the old queen is replaced as soon as a daughter queen mates and begins laying. Otherwise, she is quickly superseded in the new home.

Factors that trigger swarming

It is generally accepted that a colony of bees does not swarm until they have completed all of their brood combs, i.e., filled all available space with eggs, larvae, and brood. This generally occurs in late spring at a time when the other areas of the hive are rapidly filling with honey stores. One key trigger of the swarming instinct is when the queen has no more room to lay eggs and the hive population is becoming very congested. Under these conditions, a prime swarm may issue with the queen, resulting in a halving of the population within the hive, leaving the old colony with a large number of hatching bees. The queen who leaves finds herself in a new hive with no eggs and no larvae but lots of energetic young bees who create a new set of brood combs from scratch in a very short time.

Another important factor in swarming is the age of the queen. Those under a year in age are unlikely to swarm unless they are extremely crowded, while older queens have swarming predisposition.

Beekeepers monitor their colonies carefully in spring and watch for the appearance of queen cells, which are a dramatic signal that the colony is determined to swarm.

When a colony has decided to swarm, queen cells are produced in numbers varying to a dozen or more. When the first of these queen cells is sealed after eight days of larval feeding, a virgin queen pupates and is due to emerge seven days later. Before leaving, the worker bees fill their stomachs with honey in preparation for the creation of new honeycombs in a new home. This cargo of honey also makes swarming bees less inclined to sting. A newly issued swarm is noticeably gentle for up to 24 hours and is often capable of being handled by a beekeeper without gloves or veil.

A swarm attached to a branch

This swarm looks for shelter. A beekeeper may capture it and introduce it into a new hive, helping meet this need. Otherwise, it returns to a feral state, in which case it finds shelter in a hollow tree, excavation, abandoned chimney, or even behind shutters.

Back at the original hive, the first virgin queen to emerge from her cell immediately seeks to kill all her rival queens still waiting to emerge. Usually, however, the bees deliberately prevent her from doing this, in which case, she too leads a second swarm from the hive. Successive swarms are called 'after-swarms' or 'casts' and can be very small, often with just a thousand or so bees—as opposed to a prime swarm, which may contain as many as ten to twenty-thousand bees.

A small after-swarm has less chance of survival and may threaten the original hive's survival if the number of individuals left is unsustainable. When a hive swarms despite the beekeeper's preventative efforts, a good management practice is to give the reduced hive a couple frames of open brood with eggs. This helps replenish the hive more quickly and gives a second opportunity to raise a queen if there is a mating failure.

Each race or sub-species of honey bee has its own swarming characteristics. Italian bees are very prolific and inclined to swarm; Northern European black bees have a strong tendency to supersede their old queen without swarming. These differences are the result of differing evolutionary pressures in the regions where each sub-species evolved.

Artificial swarming

When a colony accidentally loses its queen, it is said to be "queenless". The workers realize that the queen is absent after as little as an hour, as her pheromones fade in the hive. The colony cannot survive without a fertile queen laying eggs to renew the population, so the workers select cells containing eggs aged less than three days and enlarge these cells dramatically to form "emergency queen cells". These appear similar to large peanut-like structures about an inch long that hang from the center or side of the brood combs. The developing larva in a queen cell is fed differently from an ordinary worker-bee; in addition to the normal honey and pollen, she receives a great deal of royal jelly, a special food secreted by young 'nurse bees' from the hypopharyngeal gland. This special food dramatically alters the growth and development of the larva so that, after metamorphosis and pupation, it emerges from the cell as a queen bee. The queen is the only bee in a colony which has fully developed ovaries, and she secretes a pheromone which suppresses the normal development of ovaries in all her workers.

Beekeepers use the ability of the bees to produce new queens to increase their colonies in a procedure called splitting a colony. To do this, they remove several brood combs from a healthy hive, taking care to leave the old queen behind. These combs must contain eggs or larvae less than three days old and be covered by young nurse bees, which care for the brood and keep it warm. These brood combs and attendant nurse bees are then placed into a small 'nucleus hive' with other combs containing honey and pollen. As soon as the nurse bees find themselves in this new hive and realize they have no queen, they set about constructing emergency queen cells using the eggs or larvae they have in the combs with them.

Diseases

The common agents of disease that affect adult honey bees include fungi, bacteria, protozoa, viruses, parasites, and poisons. The gross symptoms displayed by affected adult bees are very similar, whatever the cause, making it difficult for the apiarist to ascertain the causes of problems without microscopic identification of microorganisms or chemical analysis of poisons. Since 2006 colony losses from Colony Collapse Disorder have been increasing across the world although the causes of the syndrome are, as yet, unknown. In the US, commercial beekeepers have been increasing the number of hives to deal with higher rates attrition.

HOW TO GET STARTED STEP BY STEP

Starting a business involves planning, making key financial decisions and completing a series of legal activities. These 12 easy steps can help you plan, prepare and manage your business.

Step 1: Write a Business Plan

Use these tools and resources to create a business plan. This written guide will help you map out how you will start and run your business successfully.

Step 2: Get Business Assistance and Training

Take advantage of free training and counseling services, from preparing a business plan and securing financing, to expanding or relocating a business from the Small Business Administration.

Step 3: Choose a Business Location

Get advice on how to select a customer-friendly location and comply with zoning laws.

Step 4: Finance Your Business

Find government backed loans, venture capital and research grants to help you get started.

Step 5: Determine the Legal Structure of Your Business
Decide which form of ownership is best for you: sole proprietorship, partnership, Limited Liability Company (LLC), corporation, S corporation, nonprofit or cooperative.

Step 6: Register a Business Name ("Doing Business As")
Register your business name with your state government.

Step 7: Get a Tax Identification Number

Learn which tax identification number you'll need to obtain from the IRS and your state revenue agency.

Step 8: Register for State and Local Taxes

Register with your state to obtain a tax identification number, workers' compensation, unemployment and disability insurance.

Step 9: Obtain Business Licenses and Permits

Get a list of federal, state and local licenses and permits required for your business.

Step 10: Understand Employer Responsibilities

Learn the legal steps you need to take to hire employees.

Step 11: Get Equipment and Supplies

Get everything together that you'll need in order to actually operate. This includes items such as a truck, chemicals, equipment, and the various business forms such as service contracts. Once you have these things together, you can start the marketing process in order to get new customers.

Step 12: Your Marketing Plan

Coming up with your overall marketing plan, and implementing that plan. When you're just starting, it is usually best to choose one or two major marketing strategies, and work on those until you're getting a steady stream of customers. Once you've gotten good at once specific marketing avenue, then it's a good idea to move on to another one, and repeat the process. You can begin with "Zero cost marketing" and scale up once you are bringing in constant sales.

How to Write a Business Plan

Millions of people want to know what is the secret to making money. Most have come to the conclusion that it is to start a business. So how to start a business? The first thing you do to start is business is to create a business plan.

A business plan is a formal statement of a set of business goals, the reasons they are believed attainable, and the plan for reaching those goals. It may also contain background information about the organization or team attempting to reach those goals.

A professional business plan consists of ten parts.

1. Executive Summary

The executive summary is often considered the most important section of a business plan. This section briefly tells your reader where your company is, where you want to take it, and why your business idea will be successful. If you are seeking financing, the executive summary is also your first opportunity to grab a potential investor's interest.

2. Company Description

This section of your plan provides a high-level review of the different elements of your business. This is akin to an extended elevator pitch and can help readers and potential investors quickly understand the goal of your business and its unique proposition.

3. Market Analysis

The market analysis section of your plan should illustrate your industry and market knowledge as well as any of your research findings and conclusions. This section is usually presented after the company description.

4. Organization and Management

Organization and Management follows the Market Analysis. This section should include: your company's organizational structure, details about the ownership of your company, profiles of your management team, and the qualifications of your board of directors.

5. Service or Product Line

Once you've completed the Organizational and Management section of your plan, the next part of your plan is where you describe your service or product, emphasizing the benefits to potential and current customers. Focus on why your particular product will fill a need for your target customers.

6. Marketing and Sales

Once you've completed the Service or Product Line section of your plan, the next part of your plan should focus on your marketing and sales management strategy for your business.

7. Funding Request

If you are seeking funding for your business venture, use this section to outline your requirements.

8. Financial Projections

You should develop the Financial Projections section after you've analyzed the market and set clear objectives. That's when you can allocate resources efficiently. The following is a list of the critical financial statements to include in your business plan packet.

9. Marketing and Sales

Once you've completed the Service or Product Line section of your plan, the next part of your business plan should focus on your marketing and sales management strategy for your business.

10. Appendix

The Appendix should be provided to readers on an as-needed basis. In other words, it should not be included with the main body of your business plan. Your plan is your communication tool; as such, it will be seen by a lot of people. Some of the information in the business section you will not want everyone to see, but specific individuals (such as creditors) may want access to this information to make lending decisions. Therefore, it is important to have the appendix within easy reach.

How to make your business plan stand out.

One of the first steps to business planning is determining your target market and why they would want to buy from you.

For example, is the market you serve the best one for your product or service? Are the benefits of dealing with your business clear and are they aligned with customer needs? If you're unsure about the answers to any of these questions, take a step back and revisit the foundation of your business plan.

YOUR GOLDMINE OF GOVERNMENT GRANTS!

Government grants. Many people either don't believe government grants exsist or they don't think they would ever be able to get government grant money.

First lets make one thing clear. Government grant money is **YOUR MONEY**. Government money comes from taxes paid by residents of this country. Depending on what state you live in, you are paying taxes on almost everything....Property tax for your house. Property tax on your car. Taxes on the things you purchase in the mall, or at the gas station. Taxes on your gasoline, the food you buy etc.

So get yourself in the frame of mind that you are not a charity case or too proud to ask for help, because billionaire companies like GM, Big Banks and most of Corporate America is not hesitating to get their share of **YOUR MONEY**!

There are over two thousand three hundred (2,300) Federal Government Assistance Programs. Some are loans but many are formula grants and project grants. To see all of the programs available go to:

http://www.CFDA.gov

WRITING A GRANT PROPOSAL

The Basic Components of a Proposal

There are eight basic components to creating a solid proposal package:

(5) The proposal summary;

(6) Introduction of organization;

(7) The problem statement (or needs assessment);

(8) Project objectives;

(9) Project methods or design;

(10) Project evaluation;

(11) Future funding; and

(12) The project budget.

The following will provide an overview of these components.

1. The Proposal Summary: Outline of Project Goals

The proposal summary outlines the proposed project and should appear at the beginning of the proposal. It could be in the form of a cover letter or a separate page, but should definitely be brief -- no longer than two or three paragraphs. The summary would be most useful if it were prepared after the proposal has been developed in order to encompass all the key summary points necessary to communicate the objectives of the project. It is this document that becomes the cornerstone of your proposal, and the initial impression it gives will be critical to the success of your venture. In many cases, the summary will be the first part of the proposal package seen by agency officials and very possibly could be the only part of the package that is carefully reviewed before the decision is made to consider the project any further.

The applicant must select a fundable project which can be supported in view of the local need. Alternatives, in the absence of Federal support, should be pointed out. The influence of the project both during and after the project period should be explained. The consequences of the project as a result of funding should be highlighted.

2. Introduction: Presenting a Credible Applicant or Organization

The applicant should gather data about its organization from all available sources. Most proposals require a description of an applicant's organization to describe its past and present operations. Some features to consider are:

A brief biography of board members and key staff members.

The organization's goals, philosophy, track record with other grantors, and any success stories.

The data should be relevant to the goals of the Federal grantor agency and should establish the applicant's credibility.

3. The Problem Statement: Stating the Purpose at Hand

The problem statement (or needs assessment) is a key element of a proposal that makes a clear, concise, and well-supported statement of the problem to be addressed. The best way to collect information about the problem is to conduct and document both a formal and informal needs assessment for a program in the FF-4 11-08 target or service area. The information provided should be both factual and directly related to the problem addressed by the proposal. Areas to document are:

The purpose for developing the proposal.

The beneficiaries -- who are they and how will they benefit.

The social and economic costs to be affected.

The nature of the problem (provide as much hard evidence as possible).

How the applicant organization came to realize the problem exists, and what is currently being done about the problem.

The remaining alternatives available when funding has been exhausted. Explain what will happen to the project and the impending implications.

Most importantly, the specific manner through which problems might be solved. Review the resources needed, considering how they will be used and to what end.

There is a considerable body of literature on the exact assessment techniques to be used. Any local, regional, or State government planning office, or local university offering course work in planning and evaluation techniques should be able to provide excellent background references.

Types of data that may be collected include: historical, geographic, quantitative, factual, statistical, and philosophical information, as well as studies completed by colleges, and literature searches from public or university libraries.

Local colleges or universities which have a department or section related to the proposal topic may help determine if there is interest in developing a student or faculty project to conduct a needs assessment. It may be helpful to include examples of the findings for highlighting in the proposal.

4. Project Objectives: Goals and Desired Outcome

Program objectives refer to specific activities in a proposal. It is necessary to identify all objectives related to the goals to be reached, and the methods to be employed to achieve the stated objectives. Consider quantities or things measurable and refer to a problem statement and the outcome of proposed activities when developing a well-stated objective. The figures used should be verifiable. Remember, if the proposal is funded, the stated objectives will probably be used to evaluate program progress, so be realistic. There is literature available to help identify and write program objectives.

5. Program Methods and Program Design: A Plan of Action

The program design refers to how the project is expected to work and solve the stated problem. Sketch out the following:

The activities to occur along with the related resources and staff needed to operate the project(inputs).

A flow chart of the organizational features of the project. Describe how the parts interrelate, where personnel will be needed, and what they are expected to do. Identify the kinds of facilities, transportation, and support services required (throughputs).

Explain what will be achieved through 1 and 2 above (outputs); i.e., plan for measurable results. Project staff may be required to produce evidence of program performance through an examination of stated objectives during either a site visit by the Federal grantor agency and or grant reviews which may involve peer review committees.

It may be useful to devise a diagram of the program design. For example, draw a three column block. Each column is headed by one of the parts (inputs, throughputs and outputs), and on the left 11-08 FF-5 (next to the first column) specific program features should be identified (i.e., implementation, staffing, procurement, and systems development).

In the grid, specify something about the program design, for example, assume the first column is labeled inputs and the first row is labeled staff. On the grid one might specify under inputs five nurses to operate a child care unit. The throughput might be to maintain charts, counsel the children, and set up a daily routine; outputs might be to discharge 25 healthy children per week. This type of procedure will help to conceptualize both the scope and detail of the project.

Wherever possible, justify in the narrative the course of action taken. The most economical method should be used that does not compromise or sacrifice project quality. The financial expenses associated with performance of the project will later become points of negotiation with the Federal program staff. If everything is not carefully justified in writing in the proposal, after negotiation with the Federal grantor agencies, the approved project may resemble less of the original concept.

Carefully consider the pressures of the proposed implementation, that is, the time and money needed to acquire each part of the plan. A Program Evaluation and Review Technique (PERT) chart could be useful and supportive in justifying some proposals.

The remaining alternatives available when funding has been exhausted. Explain what will happen to the project and the impending implications.

Highlight the innovative features of the proposal which could be considered distinct from other proposals under consideration.

Whenever possible, use appendices to provide details, supplementary data, references, and information requiring in-depth analysis. These types of data, although supportive of the proposal, if included in the body of the design, could detract from its readability. Appendices provide the proposal reader with immediate access to details if and when clarification of an idea, sequence or conclusion is required. Time tables, work plans, schedules, activities, methodologies, legal papers, personal vitae, letters of support, and endorsements are examples of appendices.

6. Evaluation: Product and Process Analysis

The evaluation component is two-fold: (1) product evaluation; and (2) process evaluation. Product evaluation addresses results that can be attributed to the project, as well as the extent to which the project has satisfied its desired objectives. Process evaluation addresses how the project was conducted, in terms of consistency with the stated plan of action and the effectiveness of the various activities within the plan.

Most Federal agencies now require some form of program evaluation among grantees. The requirements of the proposed project should be explored carefully. Evaluations may be conducted by an internal staff member, an evaluation firm or both. The applicant should state the amount of time needed to evaluate, how the feedback will be distributed among the proposed staff, and a schedule for review and comment for this type of communication.

Evaluation designs may start at the beginning, middle or end of a project, but the applicant should specify a start-up time. It is practical to submit an evaluation design at the start of a project for two reasons:

Convincing evaluations require the collection of appropriate data before and during program operations;

If the evaluation design cannot be prepared at the outset then a critical review of the program design may be advisable.

Even if the evaluation design has to be revised as the project progresses, it is much easier and cheaper to modify a good design. If the problem is not well defined and carefully analyzed for cause and effect relationships then a good evaluation design may be difficult to achieve. Sometimes a pilot study is needed to begin the identification of facts and relationships. Often a thorough literature search may be sufficient.

Evaluation requires both coordination and agreement among program decision makers (if known). Above all, the Federal grantor agency's requirements should be highlighted in the evaluation design. Also, Federal grantor agencies may require specific evaluation techniques such as designated data formats (an existing FF-6 11-08 information collection system) or they may offer financial inducements for voluntary participation in a national evaluation study.

The applicant should ask specifically about these points. Also, consult the Criteria For Selecting Proposals section of the Catalog program description to determine the exact evaluation methods to be required for the program if funded.

7. Future Funding: Long-Term Project Planning

Describe a plan for continuation beyond the grant period, and/or the availability of other resources necessary to implement the grant. Discuss maintenance and future program funding if program is for construction activity. Account for other needed expenditures if program includes purchase of equipment.

8. The Proposal Budget: Planning the Budget

Funding levels in Federal assistance programs change yearly. It is useful to review the appropriations over the past several years to try to project future funding levels (see Financial Information section of the Catalog program description).

However, it is safer to never anticipate that the income from the grant will be the sole support for the project. This consideration should be given to the overall budget requirements, and in particular, to budget line items most subject to inflationary pressures. Restraint is important in determining inflationary cost projections (avoid padding budget line items), but attempt to anticipate possible future increases.

Some vulnerable budget areas are: utilities, rental of buildings and equipment, salary increases, food, telephones, insurance, and transportation. Budget adjustments are sometimes made after the grant award, but this can be a lengthy process. Be certain that implementation, continuation and phase-down costs can be met. Consider costs associated with leases, evaluation systems, hard/soft match requirements, audits, development, implementation and maintenance of information and accounting systems, and other long-term financial commitments.

A well-prepared budget justifies all expenses and is consistent with the proposal narrative. Some areas in need of an evaluation for consistency are:

(13) the salaries in the proposal in relation to those of the applicant organization should be similar;

(14) if new staff persons are being hired, additional space and equipment should be considered, as necessary;

(15) if the budget calls for an equipment purchase, it should be the type allowed by the grantor agency;

(16) if additional space is rented, the increase in insurance should be supported;

(17) if an indirect cost rate applies to the proposal, the division between direct and indirect costs should not be in conflict, and the aggregate budget totals should refer directly to the approved formula; and

(18) if matching costs are required, the contributions to the matching fund should be taken out of the budget unless otherwise specified in the application instructions.

It is very important to become familiar with Government-wide circular requirements. The Catalog identifies in the program description section

(as information is provided from the agencies)

the particular circulars applicable to a Federal program, and summarizes coordination of Executive Order 12372, "Intergovernmental Review of Programs" requirements in Appendix I. The applicant should thoroughly review the appropriate circulars since they are essential in determining items such as cost principles and conforming with Government guidelines for Federal domestic assistance.

General Small Business Loans:

Loan Program, SBA's most common loan program, includes financial help for businesses with special requirements.

Loan Program Eligibility

SBA provides loans to businesses; so the requirements of eligibility are based on specific aspects of the business and its principals. As such, the key factors of eligibility are based on what the business does to receive its income, the character of its ownership and where the business operates.

SBA generally does not specify what businesses are eligible. Rather, the agency outlines what businesses are not eligible. However, there are some universally applicable requirements. To be eligible for assistance, businesses must:

Operate for profit

Be small, as defined by SBA

Be engaged in, or propose to do business in, the United States or its possessions

Have reasonable invested equity

Use alternative financial resources, including personal assets, before seeking financial assistance

Be able to demonstrate a need for the loan proceeds

Use the funds for a sound business purpose

Not be delinquent on any existing debt obligations to the U.S. Government.

Ineligible Businesses

A business must be engaged in an activity SBA determines as acceptable for financial assistance from a federal provider. The following list of businesses types are not eligible for assistance because of the activities they conduct:

Financial businesses primarily engaged in the business of lending, such as banks, finance companies, payday lenders, some leasing companies and factors (pawn shops, although engaged in lending, may qualify in some circumstances)

Businesses owned by developers and landlords that do not actively use or occupy the assets acquired or improved with the loan proceeds (except when the property is leased to the business at zero profit for the property's owners)

Life insurance companies

Businesses located in a foreign country (businesses in the U.S. owned by aliens may qualify)

Businesses engaged in pyramid sale distribution plans, where a participant's primary incentive is based on the sales made by an ever-increasing number of participants.

Businesses deriving more than one-third of gross annual revenue from legal gambling activities

Businesses engaged in any illegal activity

Private clubs and businesses that limit the number of memberships for reasons other than capacity

Government-owned entities

Businesses principally engaged in teaching, instructing, counseling or indoctrinating religion or religious beliefs, whether in a religious or secular setting

Consumer and marketing cooperatives (producer cooperatives are eligible)

Loan packagers earning more than one third of their gross annual revenue from packaging SBA loans

Businesses in which the lender or CDC, or any of its associates owns an equity interest.

Businesses that present live performances of an indecent sexual nature or derive directly or indirectly more 2.5 percent of gross revenue through the sale of products or services, or the presentation of any depictions or displays, of an indecent sexual nature

Businesses primarily engaged in political or lobbying activities

Speculative businesses (such as oil exploration)

There are also eligibility factors for financial assistance based on the activities of the owners and the historical operation of the business. As such, the business cannot have been:

A business that caused the government to have incurred a loss related to a prior business debt

A business owned 20 percent or more by a person associated with a different business that caused the government to have incurred a loss related to a prior business debt

A business owned 20 percent or more by a person who is incarcerated, on probation, on parole, or has been indicted for a felony or a crime of moral depravity.

Special Considerations

Special considerations apply to some types of businesses and individuals, which include:

Franchises are eligible except when a franchiser retains power to control operations to such an extent as to equate to an employment contract; the franchisee must have the right to profit from efforts commensurate with ownership

Recreational facilities and clubs are eligible if the facilities are open to the general public, or in membership-only situations, membership is not selectively denied or restricted to any particular groups

Farms and agricultural businesses are eligible, but these applicants should first explore Farm Service Agency (FSA) programs, particularly if the applicant has a prior or existing relationship with FSA

Fishing vessels are eligible, but those seeking funds for the construction or reconditioning of vessels with a cargo capacity of five tons or more must first request financing from the National Marine Fisheries Service

Privately owned medical facilities including hospitals, clinics, emergency outpatient facilities, and medical and dental laboratories are eligible; recovery and nursing homes are also eligible, provided they are licensed by the appropriate government agency and they provide more than room and board.

An Eligible Passive Company (EPC) must use loan proceeds to acquire or lease, and/or improve or renovate, real or personal property that it leases to one or more operating companies and must not make any profit from conducting its activities.

Legal aliens are eligible; however, consideration is given to status (e.g., resident, lawful temporary resident) in determining the business' degree of risk

Probation or parole: Applications will not be accepted from firms in which a principal is currently incarcerated, on parole, on probation or is a defendant in a criminal proceeding

Loan Amounts, Fees & Interest Rates

The specific terms of SBA loans are negotiated between a borrower and an SBA-approved lender. In general, the following provisions apply to all SBA 7(a) loans.

Loan Amounts

7(a) loans have a maximum loan amount of $5 million. SBA does not set a minimum loan amount. The average 7(a) loan amount in fiscal year 2015 was $371,628.

Fees

Loans guaranteed by the SBA are assessed a guarantee fee. This fee is based on the loan's maturity and the dollar amount guaranteed, not the total loan amount.

The lender initially pays the guaranty fee and they have the option to pass that expense on to the borrower at closing. The funds to reimburse the lender can be included in the overall loan proceeds.

On loans under $150,000 made after October 1, 2013, the fees will be set at zero percent. On any loan greater than $150,000 with a maturity of one year or shorter, the fee is 0.25 percent of the guaranteed portion of the loan.

On loans with maturities of more than one year, the normal fee is 3 percent of the SBA-guaranteed portion on loans of $150,000 to $700,000, and 3.5 percent on loans of more than $700,000. There is also an additional fee of 0.25 percent on any guaranteed portion of more than $1 million.

Interest Rates

The actual interest rate for a 7(a) loan guaranteed by the SBA is negotiated between the applicant and lender and subject to the SBA maximums. Both fixed and variable interest rate structures are available.

The maximum rate is composed of two parts, a base rate and an allowable spread. There are three acceptable base rates (A prime rate published in a daily national newspaper*, London Interbank One Month Prime plus 3 percent and an SBA Peg Rate).

Lenders are allowed to add an additional spread to the base rate to arrive at the final rate. For loans with maturities of shorter than seven years, the maximum spread will be no more than 2.25 percent.

For loans with maturities of seven years or more, the maximum spread will be 2.75 percent. The spread on loans of less than $50,000 and loans processed through Express procedures have higher maximums.

All references to the prime rate refer to the base rate in effect on the first business day of the month the loan application is received by the SBA.

Percentage of Guarantee

SBA can guarantee as much as 85 percent on loans of up to $150,000 and 75 percent on loans of more than $150,000. SBA's maximum exposure amount is $3,750,000. Thus, if a business receives an SBA-guaranteed loan for $5 million, the maximum guarantee to the lender will be $3,750,000 or 75%. SBA Express loans have a maximum guarantee set at 50 percent.
7(a) Loan Application Checklist

The specific terms of SBA loans are negotiated between a borrower and an SBA-approved lender. In general, the following provisions apply to all SBA 7(a) loans.

Loan Processing Time

There are two 7(a) loan process options with different time frames. In addition to standard procedures, SBA Express processing offers an expedited turnaround.

Special Types of 7(a) Loans

SBA offers several special purpose 7(a) loans to aid businesses that have been impacted by NAFTA, provide financial assistance to Employee Stock Ownership Plans, and help implement pollution controls.

SBA Microloan Program

The Microloan program provides loans up to $50,000 to help small businesses and certain not-for-profit childcare centers start up and expand. The average microloan is about $13,000.

The U.S. Small Business Administration provides funds to specially designated intermediary lenders, which are nonprofit community-based organizations with experience in lending as well as management and technical assistance. These intermediaries administer the Microloan program for eligible borrowers.

Eligibility Requirements

Each intermediary lender has its own lending and credit requirements. Generally, intermediaries require some type of collateral as well as the personal guarantee of the business owner.

Use of Microloan Proceeds

Microloans can be used for:

- Working capital,
- Inventory or supplies
- Furniture or fixtures,
- Machinery or equipment

Proceeds from an SBA microloan cannot be used to pay existing debts or to purchase real estate.

Repayment Terms, Interest Rates, and Fees

Loan repayment terms vary according to several factors:

- Loan amount
- Planned use of funds
- Requirements determined by the intermediary lender
- Needs of the small business borrower

The maximum repayment term allowed for an SBA microloan is six years.

Interest rates vary, depending on the intermediary lender and costs to the intermediary from the U.S. Treasury. Generally, these rates will be between 8 and 13 percent.

Application Process

Microloans are available through certain nonprofit, community-based organizations that are experienced in lending and business management assistance. If you apply for SBA microloan financing, you may be required to fulfill training or planning requirements before your loan application is considered. This business training is designed to help you launch or expand your business.

Find a Microloan Provider

To apply for a Microloan, you must work with an SBA approved intermediary in your area. Approved intermediaries make all credit decisions on SBA microloans. For more information, you can contact your local SBA District Office.

Business Legal Structure

When you are starting a business, one of the first decisions you have to make is the type of business you want to create. A sole proprietorship? A corporation? A limited liability company? This decision is important, because the type of business you create determines the types of applications you’ll need to submit. You should also research liability implications for personal investments you make into your business, as well as the taxes you will need to pay. It’s important to understand each business type and select the one that is best suited for your situation and objectives. Keep in mind that you may need to contact several federal agencies, as well as your state business entity registration office.

Here is a list of the most common ways to structure a business.

An S corporation (sometimes referred to as an S Corp) is a special type of corporation created through an IRS tax election. An eligible domestic corporation can avoid double taxation (once to the corporation and again to the shareholders) by electing to be treated as an S corporation.

A partnership is a single business where two or more people share ownership.

Each partner contributes to all aspects of the business, including money, property, labor or skill. In return, each partner shares in the profits and losses of the business.

A limited liability company is a hybrid type of legal structure that provides the limited liability features of a corporation and the tax efficiencies and operational flexibility of a partnership.

The "owners" of an LLC are referred to as "members." Depending on the state, the members can consist of a single individual (one owner), two or more individuals, corporations or other LLCs.

Corporation (C Corporation)

A corporation (sometimes referred to as a C corporation) is an independent legal entity owned by shareholders. This means that the corporation itself, not the shareholders that own it, is held legally liable for the actions and debts the business incurs.

Corporations are more complex than other business structures because they tend to have costly administrative fees and complex tax and legal requirements. Because of these issues, corporations are generally suggested for established, larger companies with multiple employees.

A cooperative is a business or organization owned by and operated for the benefit of those using its services. Profits and earnings generated by the cooperative are distributed among the members, also known as user-owners.

Typically, an elected board of directors and officers run the cooperative while regular members have voting power to control the direction of the cooperative. Members can become part of the cooperative by purchasing shares, though the amount of shares they hold does not affect the weight of their vote.

Selecting The Right Business Name

Ask 500 people already in business how they decided upon their business name and you will get 500 different answers. Everyone has a story behind how they chose their own business name. Even if the business is named after their own birth name, there's a reason why this was done.

When you open a business, in a sense, you are causing a new birth to begin. This new birth was created from an idea alone by you or your associates. It will have its own bank account, it's own federal identification number, it's own credit accounts, it's own income and it's own bills. On paper, it is another individual! Just as if you were choosing a name for your own unborn child, you need to spend considerable time in deciding upon your business name.

There are several reasons why a good business name is vitally important to your business. The first obvious reason is because it is the initial identification to your customers. No one would want to do business with someone if they didn't have a company name yet.

This makes you look like an amateur who is very unreliable. Even if you call your company "Kevin's Lawn Service," a company name has been established and you are indeed a company. People will therefore feel more comfortable dealing with you.

Secondly, a business name normally is an indication as to the product or service you offer. "Mary's Typing Service," "Karate Club for Men," "Jim-Dandy Jack-of-all-Trades," "Laurie and Steve's Laundry," "Misty's Gift Boutique," and "Star 1 Publishers" are all examples of simple business names that immediately tell the customer what product you offer.

However, most people will choose the simple approach when naming their business. They use their name, their spouse's name, their children's names or a combination of these names when naming a business. The national hamburger-restaurant chain "Wendy's" was named after the owner's daughter.

However, research has proven that these "cutesy" names are not the best names to use for a business. Many experts claim that it makes the business look too "mom-and-pop-sie." However, this depends on the business. If you are selling something that demands this mood or theme to appeal to your market, it's best to use this approach.

Personally, I am inclined to name my businesses with catchy names that stick in people's heads after we have initially made contact. Names like, "Sensible Solutions," "Direct Defenders," "Moonlighters Ink," "Printer's Friend," "Strictly Class," "Collections and Treasures," and "Starlight on Twilight" are all good examples of catchy names. These types of names relate to your product or service but serve as a type of slogan for your business. This is a big help when marketing.

A friend I know owns a business called "Mint and Pepper." He grows and sells his own line of raw seasonings to people in the local area. At a get-together for small businesses, he passed out his business card. The card had a peppermint candy glued on the back and the slogan read: "Your business is worth a mint to us." This marketing concept not only got my friend noticed and remembered, but brought in several large orders for the business.

When you name a child, you may not decide upon a definite name until after they are born. You do this because a name is sometimes associated with a type of personality. When you name a business you may need to wait until you have a product or service to sell and then decide upon a business name before going into the business itself because your business name should give some clue as to what product or service you are selling.

A business named "Joe's Collections" normally wouldn't sell car parts and a business named "Charlie Horse" would not sell knitting supplies.

To generate ideas - begin looking at business signs everywhere you go. Notice which ones catch your eye and stick in your mind. Try and figure out "why" they stuck in your mind. Naturally, the business "Dominos Pizza" sticks in your mind because it is nationally known. These don't count!

Look around and notice the smaller businesses. Take your time. Within a few days you should be able to come up with a few potential business names.

Then, when you finally find a few names you really like - try reciting them to other people and get their opinion. It won't be long until your business will have the proper name that will carry it through it's life!

HINT:

Try to avoid very long names so they will fit into small display ads. Amalgamated International Enterprises can be easily presented as AIE - which is easier and shorter to spell.

Register Your Business Name

Naming your business is an important branding exercise, but if you choose to name your business as anything other than your own personal name then you'll need to register it with the appropriate authorities.

This process is known as registering your "Doing Business As" (DBA) name.

What is a "Doing Business As" Name?

A fictitious name (or assumed name, trade name or DBA name) is a business name that is different from your personal name, the names of your partners or the officially registered name of your LLC or corporation.

It's important to note that when you form a business, the legal name of the business defaults to the name of the person or entity that owns the business, unless you choose to rename it and register it as a DBA name.

For example, consider this scenario: John Smith sets up a painting business. Rather than operate under his own name, John instead chooses to name his business: "John Smith Painting". This name is considered an assumed name and John will need to register it with the appropriate local government agency. The legal name of your business is required on all government forms and applications, including your application for employer tax IDs, licenses and permits.

Do I Need a "Doing Business As" Name?

A DBA is needed in the following scenarios:

Sole Proprietors or Partnerships – If you wish to start a business under anything other than your real name, you'll need to register a DBA so that you can do business as another name.

Existing Corporations or LLCs – If your business is already set up and you want to do business under a name other than your existing corporation or LLC name, you will need to register a DBA.

Note: Not all states require the registering of fictitious business names or DBAs.

Registering your DBA is done either with your county clerk's office or with your state government, depending on where your business is located. There are a few states that do not require the registering of fictitious business names.

Business Tax Advantages

Every year, several thousand people develop an interest in "going into business." Many of these people have an idea, a product or a service they hope to promote into an income producing business
which they can operate from their homes.

If you are one of these people, here are some practical thoughts to consider before hanging out the "Open for Business" sign.

In areas zoned "Residential Only," your proposed business could be illegal. In many areas, zoning restrictions rule out home businesses involving the coming and going of many customers, clients or employees. Many businesses that sell or even store
anything for sale on the premises also fall into this category.

Be sure to check with your local zoning office to see how the ordinances in your particular area may affect your business plans. You may need a special permit to operate your business from your home; and you may find that making small changes in your plan will put you into the position of meeting zoning
standards.

Many communities grant home occupation permits for businesses involve typing, sewing, and teaching, but turn thumbs down on requests from photographers, interior decorators and home improvement businesses to be run from the home.

And often, even if you are permitted to use your home for a given business, there will be restrictions that you may need to take into consideration. By all means, work with your zoning people, and save yourself time, trouble and dollars.

One of the requirements imposed might be off street parking for your customers or patrons. And, signs are generally forbidden in residential districts. If you teach, there is almost always a limit on the number of students you may have at any one time.

Obtaining zoning approval for your business, then, could be as simple as filling out an application, or it could involve a public hearing. The important points the zoning officials will consider will center around how your business will affect the neighborhood. Will it increase the traffic noticeably on your street? Will there be a substantial increase in noise? And how will your neighbors feel about this business alongside their homes?

To repeat, check into the zoning restrictions, and then check again to determine if you will need a city license. If you're selling something, you may need a vendor's license, and be required to collect sales taxes on your transactions. The sale tax requirement would result in the need for careful record keeping.

Licensing can be an involved process, and depending upon the type of business, it could even involve the inspection of your home to determine if it meets with local health and building and fire codes. Should this be the case, you will need to bring your facilities up to the local standards. Usually this will involve some simple repairs or adjustments that you can either do personally, or hire out to a handyman at a nominal cost.

Still more items to consider: Will your homeowner's insurance cover the property and liability in your new business? This must definitely be resolved, so be sure to talk it over with your insurance agent.

Tax deductions, which were once one of the beauties of engaging in a home business, are not what they once were. To be eligible for business related deductions today, you must use that part of your home claimed EXCLUSIVELY AND REGULARLY as either the principal location of your business, or place reserved to meet patients, clients or customers.

An interesting case in point: if you use your den or a spare bedroom as the principal place of business, working there from 8:00 to 5:00 every day, but permit your children to watch TV in that room during evening hours, the IRS dictates that you cannot claim a deduction for that room as your office or place of business.

There are, however, a couple of exceptions to the "exclusive use" rule. One is the storage on inventory in your home, where your home is the location of your trade or business, and your trade or business is the selling of products at retail or wholesale.

According to the IRS, such storage space must be used on a REGULAR Basis, and be separately identifiable space.

Another exception applies to daycare services that are provided for children, the elderly, or physically or mentally handicapped. This exception applies only if the owner of the facility complies with the state laws for licensing.

To be eligible for business deductions, your business must be an activity undertaken with the intent of making profit. It's presumed you meet this requirement if your business makes a profit in any two years of a five-year period.

Once you are this far along, you can deduct business expenses such as supplies, subscriptions to professional journals, and an allowance for the business use of your car or truck. You can also
claim deductions for home related business expenses such as utilities, and in some cases, even a new paint job for your home.

The IRS is going to treat the part of your home you use for business as though it were a separate piece of property. This means that you'll have to keep good records and take care not to mix business and personal matters. No specific method of record
keeping is required, but your records must clearly justify and deductions you claim.

You can begin by calculating what percentage of the house is used for business, Either by number of rooms or by area in square footage. Thus, if you use one of the five rooms for your business, the business portion is 20 percent. If you run your business out of a room that's 10 by 12 feet, and the total area of your home is 1,200 square feet, the business space factor is 10 percent.

An extra computation is required if your business is a home day care center. This is one of the exempted activities in which the exclusive use rule doesn't apply. Check with your tax preparer and the IRS for an exact determination.

If you're a renter, you can deduct the part of your rent which is attributable to the business share of your house or apartment. Homeowners can take a deduction based on the depreciation of the business portion of their house.

There is a limit to the amount you can deduct. This is the amount equal to the gross income generated by the business, minus those home expenses you could deduct even if you weren't operating a business from your home. As an example, real estate taxes and mortgage interest are deductible regardless of any business Activity in your home, so you must subtract from your business Gross income the percentage that's allocable to the business portion of your home. You thus arrive at the maximum amount for home-related business deductions.

If you are self-employed, you claim your business deductions on SCHEDULE C, PROFIT(or LOSS) for BUSINESS OR PROFESSION. The IRS emphasizes that claiming business-at-home deductions does not automatically trigger an audit on your tax return.

Even so, it is always wise to keep meticulously within the proper guidelines, and of course keep detailed records if you claim business related expenses when you are working out of your home. You should discuss this aspect of your operation with your tax preparer or a person qualified in the field of small business tax requirements.

If your business earnings aren't subject to withholding tax, and your estimated federal taxes are $100 or more, you'll probably be filing a Declaration of Estimated Tax, Form 1040 ES.

To complete this form, you will have to estimate your income for the coming year and also make a computation of the income tax and self-employed tax you will owe.

The self-employment taxes pay for Social Security coverage. If you have a salaried job covered by Social Security, the self-employment tax applies only to that amount of your home business income that, when added to your salary, reaches the current ceiling. When you file your Form 1040-ES, which is due April 15, you must make the first of four equal installment payments on your estimated tax bill.

Another good way to trim taxes is by setting up a Keogh plan or an Individual Retirement Account. With either of these, you can shelter some of your home business income from taxes by investing it for your retirement.

HOW TO HIRE YOUR EMPLOYEE'S

If your business is booming, but you are struggling to keep up, perhaps it's time to hire some help.

The eight steps below can help you start the hiring process and ensure you are compliant with key federal and state regulations.

Step 1. Obtain an Employer Identification Number (EIN)

Before hiring your first employee, you need to get an employment identification number (EIN) from the U.S. Internal Revenue Service. The EIN is often referred to as an Employer Tax ID or as Form SS-4. The EIN is necessary for reporting taxes and other documents to the IRS. In addition, the EIN is necessary when reporting information about your employees to state agencies. Apply for EIN online or contact the IRS at 1-800-829-4933.

Step 2. Set up Records for Withholding Taxes

According to the IRS, you must keep records of employment taxes for at least four years. Keeping good records can also help you monitor the progress of your business, prepare financial statements, identify sources of receipts, keep track of deductible expenses, prepare your tax returns, and support items reported on tax returns.

Below are three types of withholding taxes you need for your business:

Federal Income Tax Withholding

Every employee must provide an employer with a signed withholding exemption certificate (Form W-4) on or before the date of employment. The employer must then submit Form W-4 to the IRS. For specific information, read the IRS' Employer's Tax Guide.

Federal Wage and Tax Statement

Every year, employers must report to the federal government wages paid and taxes withheld for each employee. This report is filed using Form W-2, wage and tax statement. Employers must complete a W-2 form for each employee who they pay a salary, wage or other compensation.

Employers must send Copy A of W-2 forms to the Social Security Administration by the last day of February to report wages and taxes of your employees for the previous calendar year. In addition, employers should send copies of W-2 forms to their employees by Jan. 31 of the year following the reporting period. Visit SSA.gov/employer for more information.

State Taxes

Depending on the state where your employees are located, you may be required to withhold state income taxes. Visit the state and local tax page for more information.

Step 3. Employee Eligibility Verification

Federal law requires employers to verify an employee's eligibility to work in the United States. Within three days of hire, employers must complete Form I-9, employment eligibility verification, which requires employers to examine documents to confirm the employee's citizenship or eligibility to work in the U.S. Employers can only request documentation specified on the I-9 form.

Employers do not need to submit the I-9 form with the federal government but are required to keep them on file for three years after the date of hire or one year after the date of the employee's termination, whichever is later.

Employers can use information taken from the Form I-9 to electronically verify the employment eligibility of newly hired employees by registering with E-Verify.

Visit the U.S. Immigration and Customs Enforcement agency's I-9 website to download the form and find more information.

Step 4. Register with Your State's New Hire Reporting Program

All employers are required to report newly hired and re-hired employees to a state directory within 20 days of their hire or rehire date.

Step 5. Obtain Workers' Compensation Insurance

All businesses with employees are required to carry workers' compensation insurance coverage through a commercial carrier, on a self-insured basis or through their state's Workers' Compensation Insurance program.

Step 6. Post Required Notices

Employers are required to display certain posters in the workplace that inform employees of their rights and employer responsibilities under labor laws. Visit the Workplace Posters page for specific federal and state posters you'll need for your business.

Step 7. File Your Taxes

Generally, employers who pay wages subject to income tax withholding, Social Security and Medicare taxes must file IRS Form 941, Employer's Quarterly Federal Tax Return. For more information, visit IRS.gov.

New and existing employers should consult the IRS Employer's Tax Guide to understand all their federal tax filing requirements.

Step 8. Get Organized and Keep Yourself Informed

Being a good employer doesn't stop with fulfilling your various tax and reporting obligations. Maintaining a healthy and fair workplace, providing benefits and keeping employees informed about your company's policies are key to your business' success. Here are some additional steps you should take after you've hired your first employee:

Set up Recordkeeping

In addition to requirements for keeping payroll records of your employees for tax purposes, certain federal employment laws also require you to keep records about your employees. The following sites provide more information about federal reporting requirements:

Tax Recordkeeping Guidance

Labor Recordkeeping Requirements

Occupational Safety and Health Act Compliance

Employment Law Guide (employee benefits chapter)

Apply Standards that Protect Employee Rights

Complying with standards for employee rights in regards to equal opportunity and fair labor standards is a requirement. Following statutes and regulations for minimum wage, overtime, and child labor will help you avoid error and a lawsuit. See the Department of Labor's Employment Law Guide for up-to-date information on these statutes and regulations.

Also, visit the Equal Employment Opportunity Commission and Fair Labor Standards Act.

Web Wholesale Resource Rolodex

As of the writting of this book all, of the companies below, website is up and have an active business. From time to time companies go out of business or change their web address. So, instead of just giving you just 1 source I give you plenty to choose from.

Beekeeping Supplies

http://www.mannlakeltd.com/

https://www.kelleybees.com/

https://www.beethinking.com/

http://beesource.com/

http://millerbeesupply.com/catalog/

http://westernbee.com/

http://www.golden-bee.com/

http://www.worldofbeekeeping.com/free-kit/

http://apiarybeekeepingsupplies.com/

TRANSPORTATION

Used Trucks/CARS Online

http://gsaauctions.gov/gsaauctions/gsaauctions/

http://www.ebay.com/motors

http://www.uhaul.com/TruckSales/

http://www.usedtrucks.ryder.com/vehicle/VehicleSearch.aspx?VehicleTypeId=1&VehicleGroupId=3

http://www.penskeusedtrucks.com/truck-types/light-and-medium-duty/

Parts

http://www.truckchamp.com/

http://www.autopartswarehouse.com/

Bikes & Motorcycles

http://gsaauctions.gov/gsaauctions/aucindx/

http://www.bikesdirect.com/products/used-bikes/?gclid=CLCF0vaDm7kCFYtDMgodzW0AXQ

http://www.overstock.com/Sports-Toys/Cycling/450/cat.html

http://www.nashbar.com/bikes/TopCategories_10053_http://www.nashbar.com/bikes/TopCategories_10053_10052_-110052_-1

http://www.bti-usa.com/

http://evosales.com/

COMPUTERS/Office Equipment

http://www.wtsmedia.com/

http://www.laptopplaza.com/

http://www.outletpc.com/

Computer Tool Kits

http://www.dhgate.com/wholesale/computer+repair+tools.html

http://www.aliexpress.com/wholesale/wholesale-repair-computer-tool.html

http://wholesalecomputercables.com/Computer-Repair-Tool-Kit/M/B00006OXGZ.htm

http://www.tigerdirect.com/applications/category/category_tlc.asp?CatId=47&name=Computer%20Tools

Computer Parts

http://www.laptopuniverse.com/

http://www.sabcal.com/

other

http://www.nearbyexpress.com/

http://www.commercialbargains.co

http://www.getpaid2workfromhome.com

http://www.boyerblog.com/success-tools

American merchandise liquidators

http://www.amlinc.com/

the closeout club

http://www.thecloseoutclub.com/

RJ discount sales

http://www.rjsks.com/

St louis wholesale

http://www.stlouiswholesale.com/

Wholesale Electronics

http://www.weisd.com/

http://www.anawholesale.com/

office wholesale

http://www.1-computerdesks.com/

1aaa wholesale merchandise

http://www.1aaawholesalemerchandise.com/

big lots wholesale

http://www.biglotswholesale.com/

More Business Resources

1. http://www.sba.gov/content/starting-green-business

home based businesses

2. http://www.sba.gov/content/home-based-business

3. online businesses

http://www.sba.gov/content/setting-online-business

4. self employed and independent contractors

http://www.sba.gov/content/self-employed-independent-contractors

5. minority owned businesses

http://www.sba.gov/content/minority-owned-businesses

6. veteran owned businesses

http://www.sba.gov/content/veteran-service-disabled-veteran-owned

7. woman owned businesses

http://www.sba.gov/content/women-owned-businesses

8. people with disabilities

http://www.sba.gov/content/people-with-disabilities

9. young entrepreneurs

http://www.sba.gov/content/young-entrepreneurs

ZERO COST MARKETING

The web-RESOURCE guide has plenty of web sites for you to find products at huge dicounts. Below are a few steps to market those products using
ZERO COST INTERNET MARKETING stratigies.

While there are many ways to market we are only going focus on ZERO COST MARKETING. You are starting up. You can always go for the more expensive ways of marketing after your business is producing income.

FREE WEB HOSTING

Get a free web site. You can get a free web site at weebly.com or wix.com. Or just type "free web hosting" in a google, bing or yahoo search engine.

Free web hosting is something you can use for a varitey or reasons. However many free web hosting sites add an extention to the name of you web address that lets everyone know you are using their services. For this reason you eventually want to scale up once you start making income.

LOW COST PAID WEB HOSTING

Free is nice, but you when you need to expand your business it is best to go with a paid web hosting service. There are several that give you good value for under $10.00 a month.

1. Yahoo small business

2. Intuit.com

3. ipage.com

4. Hostgator.com

5. Godaddy.com

Yahoo small business allows for unlimited web pages and is probably the best overall value, but they require a years payment up front. Intuit allows for monthly payments.

For free ecommerce on your web site, open up a Paypal account and get the HTML code for payment buttons for free. Then put those buttons on your web site.

Step by Step basic zero cost web site traffic instructions

Step 1 zero cost internet marketing

Now that your web site is up and running you should register it with at least the top 3 search engines. 1. Google 2. Bing 3. Yahoo.

Step 2 zero cost internet marketing

Write and submit a press release. Google "free press release sites" for press release sites that will allow you to summit press releases for free. I you do not know how to write a press release go to www.fiverr.com and sub-contract the work out for only $5.00 !!!

Step 3 zero cost internet marketing

Write and submit articles to article marketing web sites like ezinearticles.com.

Step 4 zero cost internet marketing

Create and submit videos to video sharing sites like dailymotion.com or youtube.com. Make sure to include a hyperlink to your website in the description of your videos.

Step 5 zero cost internet marketing

Submit your web site to dmoz.org. This is a huge open directory that many smaller search engines go to get web sites for their database.

Getting Motivated to Start Your Business

The other day I was driving down a street in a commercial district and noticed a moving van. Two "blue collar workers" were busy loading office furniture in a company-owned vehicle. It was obvious that they were employees of this major company and I surmised they were making about $12 per hour.

I then thought about the normal lives these men probably lived. They had to punch a time-clock every day. If they were sick, they had to report to a boss and get permission to stay home. They had to depend on the company to pay them a weekly salary. They got paid the same amount of money every week and were lucky to get a raise every year or so.

Because of being controlled by a time clock at work, they naturally arranged their lives in the same fashion. They came home at the same time, ate dinner at the same time, looked forward to Friday for 2 days of rest but ended up cramming all their neglected responsibilities from the previous week into those 2 days.

The entire human race consists of two major groups of people: (1) Leaders and (2) Followers. Leaders have a built-in knack to not be happy with the normal flow of existence. Leaders are continually striving for a way out of this rat-race because they have a human characteristic of wanting to lead instead of follow.

But a good many of these leaders don't have a lot of money because they have been working for an employer all their lives. They recognize that they will never achieve the level of success they desire working for someone else. But they can't just leave their job and survive on their own. How could they pay the rent? Put their children through college? Buy the groceries? Pay Visa and Mastercard? With all these fears sitting in the leader will often exist as a follower because he or she doesn't believe they have a choice.

But they do! In fact, the answer is right under their noses. Allow me to explain . . .

Let's take the guy working for the moving company that I saw when I was driving down the street. He could offer the same service on weekends through word-of-mouth advertising. By placing a simple classified ad under Services Offered in the local newspaper, he could pick up a couple jobs a month and bring in an extra income.

Or how about the lady that just had a baby and wants to stay at home with it. Her maternity leave from her employer, only allows her 6 weeks. If she doesn't go back to work then she will either lose her job, her income or both. If her husband doesn't bring in enough money to support her and the baby she doesn't think she has a choice. The new mother will sacrifice money for her child.

But if this lady wants to stay at home with her child, why doesn't she start a home day-care center? That way, she would still make money and be able to be with her new child at the same time. Good for the child. Good for the mother.

Good for the family unit. Good for other working mothers who can trust a "mother-run" day care center versus a commercial one. Plus since the day care center is in this mother's home, she can charge 40% to 50% less than commercial day care centers and probably make more money compared to her old job.

Too often, people who want to break out of the mold and start their own business will seek for products and services they know absolutely nothing about. Someone told them they could make a lot of money doing this and doing that. But the truth is that it will take anyone longer to make money with a product or service that they have to learn. In fact, this learning period could take a year or more.

The person could easily be discouraged about a small business if it doesn't make any money by then.

So, if you are considering starting a small business; entertain the possibility of starting one based on the skills you already possess.

BONUS MATERIAL: CREDIT REPAIR

Credit can sometimes be the lifeblood of a business, but today many simply don't know their rights when it comes to credit.

The information below from the FTC (Federal Trade Commission) can help you to get a FREE CREDIT REPORT and begin to correct or remove any blemishes on your credit.

The Fair Credit Reporting Act (FCRA) requires each of the nationwide credit reporting companies — Equifax, Experian, and TransUnion — to provide you with a free copy of your credit report, at your request, once every 12 months. The FCRA promotes the accuracy and privacy of information in the files of the nation's credit reporting companies. The Federal Trade Commission (FTC), the nation's consumer protection agency, enforces the FCRA with respect to credit reporting companies.

A credit report includes information on where you live, how you pay your bills, and whether you've been sued or have filed for bankruptcy. Nationwide credit reporting companies sell the information in your report to creditors, insurers, employers, and other businesses that use it to evaluate your applications for credit, insurance, employment, or renting a home.

Here are the details about your rights under the FCRA, which established the free annual credit report program.

Q: How do I order my free report?

The three nationwide credit reporting companies have set up a central website, a toll-free telephone number, and a mailing address through which you can order your free annual report.

To order, visit annualcreditreport.com, call 1-877-322-8228. Or complete the Annual Credit Report https://www.consumer.ftc.gov/articles/pdf-0093-annual-report-request-form.pdfRequest Form and mail it to: Annual Credit Report Request Service, P.O. Box 105281, Atlanta, GA 30348-5281. Do not contact the three nationwide credit reporting companies individually. They are providing free annual credit reports only through annualcreditreport.com, 1-877-322-8228 or mailing to Annual Credit Report Request Service.

You may order your reports from each of the three nationwide credit reporting companies at the same time, or you can order your report from each of the companies one at a time. The law allows you to order one free copy of your report from each of the nationwide credit reporting companies every 12 months.

A Warning About "Imposter" Websites

Only one website is authorized to fill orders for the free annual credit report you are entitled to under law — annualcreditreport.com. Other websites that claim to offer "free credit reports," "free credit scores," or "free credit monitoring" are not part of the legally mandated free annual credit report program.

In some cases, the "free" product comes with strings attached. For example, some sites sign you up for a supposedly "free" service that converts to one you have to pay for after a trial period. If you don't cancel during the trial period, you may be unwittingly agreeing to let the company start charging fees to your credit card.

Some "imposter" sites use terms like "free report" in their names; others have URLs that purposely misspell annualcreditreport.com in the hope that you will mistype the name of the official site. Some of these "imposter" sites direct you to other sites that try to sell you something or collect your personal information.

Annualcreditreport.com and the nationwide credit reporting companies will not send you an email asking for your personal information. If you get an email, see a pop-up ad, or get a phone call from someone claiming to be from annualcreditreport.com or any of the three nationwide credit reporting companies, do not reply or click on any link in the message. It's probably a scam. Forward any such email to the FTC at spam@uce.gov.

Q: What information do I need to provide to get my free report?

A: You need to provide your name, address, Social Security number, and date of birth. If you have moved in the last two years, you may have to provide your previous address. To maintain the security of your file, each nationwide credit reporting company may ask you for some information that only you would know, like the amount of your monthly mortgage payment.

Each company may ask you for different information because the information each has in your file may come from different sources.

Q: Why do I want a copy of my credit report?

A: Your credit report has information that affects whether you can get a loan — and how much you will have to pay to borrow money. You want a copy of your credit report to:

- make sure the information is accurate, complete, and up-to-date before you apply for a loan for a major purchase like a house or car, buy insurance, or apply for a job.
- help guard against identity theft. That's when someone uses your personal information — like your name, your Social Security number, or your credit card number — to commit fraud. Identity thieves may use your information to open a new credit card account in your name. Then, when they don't pay the bills, the delinquent account is reported on your credit report. Inaccurate information like that could affect your ability to get credit, insurance, or even a job.

Q: How long does it take to get my report after I order it?

A: If you request your report online at annualcreditreport.com, you should be able to access it immediately. If you order your report by calling toll-free 1-877-322-8228, your report will be processed and mailed to you within 15 days. If you order your report by mail using the Annual Credit Report Request Form, your request will be processed and mailed to you within 15 days of receipt.

Whether you order your report online, by phone, or by mail, it may take longer to receive your report if the nationwide credit reporting company needs more information to verify your identity.

Q: Are there any other situations where I might be eligible for a free report?

A: Under federal law, you're entitled to a free report if a company takes adverse action against you, such as denying your application for credit, insurance, or employment, and you ask for your report within 60 days of receiving notice of the action. The notice will give you the name, address, and phone number of the credit reporting company.

You're also entitled to one free report a year if you're unemployed and plan to look for a job within 60 days; if you're on welfare; or if your report is inaccurate because of fraud, including identity theft. Otherwise, a credit reporting company may charge you a reasonable amount for another copy of your report within a 12-month period.

To buy a copy of your report, contact:

- Equifax:1-800-685-1111; equifax.com
- Experian: 1-888-397-3742; experian.com
- TransUnion: 1-800-916-8800; transunion.com

Q: Should I order a report from each of the three nationwide credit reporting companies?

A: It's up to you. Because nationwide credit reporting companies get their information from different sources, the information in your report from one company may not reflect all, or the same, information in your reports from the other two companies. That's not to say that the information in any of your reports is necessarily inaccurate; it just may be different.

Q: Should I order my reports from all three of the nationwide credit reporting companies at the same time?

A: You may order one, two, or all three reports at the same time, or you may stagger your requests. It's your choice. Some financial advisors say staggering your requests during a 12-month period may be a good way to keep an eye on the accuracy and completeness of the information in your reports.

Q: What if I find errors — either inaccuracies or incomplete information — in my credit report?

A: Under the FCRA, both the credit report ing company and the information provider (that is, the person, company, or organization that provides information about you to a consumer reporting company) are responsible for correcting inaccurate or incomplete information in your report. To take full advantage of your rights under this law, contact the credit reporting company and the information provider.

1. Tell the credit reporting company, in writing, what information you think is inaccurate.

Credit reporting companies must investigate the items in question — usually within 30 days — unless they consider your dispute frivolous. They also must forward all the relevant data you provide about the inaccuracy to the organization that provided the information. After the information provider receives notice of a dispute from the credit reporting company, it must investigate, review the relevant information, and report the results back to the credit reporting company. If the information provider finds the disputed information is inaccurate, it must notify all three nationwide credit reporting companies so they can correct the information in your file.

When the investigation is complete, the credit reporting company must give you the written results and a free copy of your report if the dispute results in a change. (This free report does not count as your annual free report.) If an item is changed or deleted, the credit reporting company cannot put the disputed information back in your file unless the information provider verifies that it is accurate and complete. The credit reporting company also must send you written notice that includes the name, address, and phone number of the information provider.

2. Tell the creditor or other information provider in writing that you dispute an item. Many providers specify an address for disputes. If the provider reports the item to a credit reporting company, it must include a notice of your dispute. And if you are correct — that is, if the information is found to be inaccurate — the information provider may not report it again.

Q: What can I do if the credit reporting company or information provider won't correct the information I dispute?

A: If an investigation doesn't resolve your dispute with the credit reporting company, you can ask that a statement of the dispute be included in your file and in future reports. You also can ask the credit reporting company to provide your state ment to anyone who received a copy of your report in the recent past. You can expect to pay a fee for this service.

If you tell the information provider that you dispute an item, a notice of your dispute must be included any time the information provider reports the item to a credit reporting company.

Q: How long can a credit reporting company report negative information?

A: A credit reporting company can report most accurate negative information for seven years and bankruptcy information for 10 years. There is no time limit on reporting information about criminal convictions; information reported in response to your application for a job that pays more than $75,000 a year; and information reported because you've applied for more than $150,000 worth of credit or life insurance. Information about a lawsuit or an unpaid judgment against you can be reported for seven years or until the statute of limitations runs out, whichever is longer.

Q: Can anyone else get a copy of my credit report?

A: The FCRA specifies who can access your credit report. Creditors, insurers, employers, and other businesses that use the information in your report to evaluate your applications for credit, insurance, employment, or renting a home are among those that have a legal right to access your report.

Q: Can my employer get my credit report?

A: Your employer can get a copy of your credit report only if you agree. A credit reporting company may not provide information about you to your employer, or to a prospective employer, without your written consent.

For More Information

The FTC works for the consumer to prevent fraudulent, deceptive, and unfair business practices in the marketplace and to provide information to help consumers spot,stop, and avoid them. To file a complaint, visit ftc.gov/complaint or call 1-877-FTC-HELP (1-877-382-4357).

Get Our Video Training Program at:

(Zero Cost Internet Marketing complete 142 video series)

http://goo.gl/gQnSo4

Massive Money for Real Estate Investing

http://www.BrianSMahoney.com

Made in the USA
Middletown, DE
05 January 2024

47320588R00070